IT'S TIME TO LEARN ABOUT IT'S TIME TO LEARN ABOUT HOW TO PASS COLLEGE ALGEBRA

It's Time to Learn about It's Time to Learn about How to Pass College Algebra

Walter the Educator

Silent King Books
A WhichHead Entertainment Imprint

Copyright © 2024 by Walter the Educator

All rights reserved. No part of this book may be reproduced in any manner whatsoever without written permission except in the case of brief quotations embodied in critical articles and reviews.

First Printing, 2024

Disclaimer

The author and publisher offer this information without warranties expressed or implied. No matter the grounds, neither the author nor the publisher will be accountable for any losses, injuries, or other damages caused by the reader's use of this book. Your use of this book acknowledges an understanding and acceptance of this disclaimer.

It's Time to Learn about It's Time to Learn about How to Pass College Algebra is a collectible little learning book by Walter the Educator that belongs to the Little Learning Books Series. Collect them all and more books at WaltertheEducator.com

IT'S TIME TO LEARN ABOUT HOW TO PASS COLLEGE ALGEBRA

INTRO

College algebra is a foundational course that serves as a gateway to more advanced mathematical and scientific studies. Whether you're pursuing a degree in science, technology, engineering, mathematics (STEM), or even in the social sciences and business, mastering algebra is essential. For many students, college algebra can feel intimidating and difficult, but with the right approach and strategies, anyone can successfully pass the course. This little book aims to provide a thorough guide on how to pass college algebra, offering practical tips, key mathematical concepts, effective study habits, and advice on overcoming common challenges.

It's Time to Learn about How to Pass College Algebra

Understanding the Importance of College Algebra

Algebra is often considered the language of mathematics. It involves the study of mathematical symbols and the rules for manipulating these symbols, which are used to express relationships and solve problems. College algebra builds upon the foundational concepts learned in high school but takes them to a more advanced level. The skills you develop in this course are crucial for success in many fields, as they help improve your analytical thinking, problem-solving abilities, and logical reasoning.

In addition to its academic significance, algebra is also applicable in everyday life. Whether you're budgeting, calculating interest rates, or analyzing data, algebraic thinking comes into play. Therefore, developing a strong understanding of college algebra will not only help you pass the course but also prepare you for real-world situations.

It's Time to Learn about How to Pass College Algebra

Building a Strong Foundation: Key Algebra Concepts

Before diving into advanced topics, it's crucial to ensure you have a solid grasp of the basic algebraic concepts. If you're rusty on any of these, it's worth taking the time to review them thoroughly. Below are the core concepts you'll need to master in college algebra:

It's Time to Learn about How to Pass College Algebra

1. **Numbers and Operations:**
   - **Real Numbers:** Understanding the properties of real numbers (whole numbers, integers, rational and irrational numbers) is fundamental. You'll work with these numbers throughout the course.

- **Order of Operations:** The order in which operations (addition, subtraction, multiplication, division) are performed is crucial in algebra. Remember the acronym PEMDAS (Parentheses, Exponents, Multiplication/Division, Addition/Subtraction) to guide you.

2. **Equations and Inequalities:**
   - **Linear Equations:** These are equations have standard form ax+by=c. Solving linear equations involves isolating the variable on one side of the equation.

- **Inequalities:** There are four types of inequalities: $>$, $<$, $\geq$, and $\leq$. Solving these is similar to solving equations but requires special attention when multiplying or dividing by negative numbers (as this reverses the inequality sign).

3. Functions:

- **Definition of a Function:** A function is a relationship where each input has exactly one output. Understanding how to work with functions, identify their domain and range, and manipulate them is a crucial part of algebra.

- **Linear, Quadratic, and Exponential Functions:** These are specific types of functions you'll encounter frequently in college algebra. Knowing how to graph and analyze their behavior is essential.

4. **Polynomials and Factoring:**
   - **Polynomials:** You'll need to know how to add, subtract, multiply, and divide polynomials.

- **Factoring:** Factoring polynomials is an important technique used to simplify expressions and solve polynomial equations. Common methods include factoring out the greatest common factor (GCF), factoring trinomials, and using the difference of squares.

5. **Rational Expressions and Equations:**
   - **Rational Expressions:** These are fractions where the numerator and/or denominator is a polynomial. Simplifying, multiplying, dividing, and solving rational expressions are important skills to master.

- **Rational Equations:** Solving rational equations involves finding a common denominator and ensuring that the solution doesn't result in division by zero.

6. **Exponents and Radicals:**
   - **Exponent Rules:** Understanding how to manipulate exponents, such as the power of a product and the power of a quotient, is crucial for solving more complex algebraic problems.

- **Radicals:** Radicals, or square roots, can be simplified and manipulated similarly to exponents. Knowing how to simplify radical expressions is important for solving equations involving roots.

7. Quadratic Equations:
- **Solving Quadratics:** Quadratic equations can be solved using methods such as factoring, completing the square, or the quadratic formula.

- **Graphing Quadratics:** The graph of a quadratic function is a parabola. Understanding how to find the vertex, axis of symmetry, and direction of the parabola is essential for interpreting quadratic equations.

Developing Effective Study Habits for College Algebra

To pass college algebra, you'll need more than just a basic understanding of the concepts. Consistent study habits and a proactive approach to learning are key. Here are some strategies to help you stay on top of the material and improve your performance:

1. **Attend All Lectures and Take Good Notes:**
   - Attending class is crucial because algebra concepts build on one another. Missing a lecture can leave you with gaps in your understanding that may affect your performance later on.
   - Take detailed notes during lectures. Write down key formulas, examples, and explanations provided by the instructor. If something is unclear, mark it for review later.

2. **Review Your Notes Regularly:**
   - Don't wait until the night before an exam to review your notes. Go over them daily to reinforce what you've learned in class. Reviewing material consistently will help you retain information and identify areas where you need further practice.

3. **Practice, Practice, Practice:**
   - Algebra is a subject that requires active practice. Simply reading your textbook won't be enough. Work through problems regularly to apply the concepts you've learned.
   - Use a variety of resources for practice, including your textbook, online tutorials, and problem sets provided by your instructor. The more problems you solve, the more familiar you'll become with different types of questions.

4. **Understand, Don't Memorize:**
   - While it's important to remember key formulas and rules, algebra is not about memorization. Focus on understanding why certain rules work and how to apply them in different contexts. This will help you approach unfamiliar problems with confidence.

5. **Ask Questions and Seek Help When Needed:**
    - Don't be afraid to ask questions if you don't understand something. Whether in class, during office hours, or in a study group, seek clarification on confusing topics before they become larger issues.
    - If you're struggling, consider working with a tutor. Many colleges offer free tutoring services for algebra students, and online tutoring platforms can also provide personalized help.

6. **Form a Study Group:**
    - Studying with classmates can be beneficial for learning algebra. In a study group, you can discuss challenging problems, share insights, and explain concepts to one another. Teaching a concept to someone else is one of the best ways to reinforce your own understanding.

7. **Use Online Resources:**
    - There are many online resources available for college algebra, including video tutorials, interactive practice problems, and forums where you can ask questions. Websites like Khan Academy, Purplemath, and Mathway offer free resources that can supplement your learning.

8. **Take Advantage of Office Hours:**
   - Your professor's office hours are a valuable opportunity to get one-on-one help with difficult topics. Use this time to ask specific questions, review challenging problems, or seek advice on how to improve your performance in the course.

Preparing for Exams: Strategies for Success

Exams in college algebra can be challenging, but with the right preparation, you can approach them with confidence. Here are some strategies to help you succeed on your exams:

1. **Start Studying Early:**
   - Don't wait until the last minute to start studying for your exam. Begin reviewing material at least a week in advance. This will give you enough time to go over all the key topics, work through practice problems, and identify any areas where you need extra help.

2. **Create a Study Schedule:**
   - Break down your study sessions into manageable chunks. Focus on one or two topics per session, and make sure to allocate time for review and practice problems. A study schedule will help you stay organized and ensure that you cover all the material before the exam.

3. **Practice with Past Exams:**
   - If your instructor provides past exams or practice exams, use them to your advantage. Working through these problems under timed conditions can help you get a feel for the types of questions that will be asked and the format of the exam.

4. **Review Key Formulas and Concepts:**
    - Make sure you know all the key formulas, rules, and concepts that will be covered on the exam. Create a cheat sheet (if allowed) or flashcards to help you review these quickly.
    - Focus on understanding how to apply formulas rather than just memorizing them. Many exam problems will require you to use formulas in creative ways.

5. **Work on Weak Areas:**
   - Identify the areas where you struggle the most, and spend extra time reviewing and practicing those topics. Don't neglect your weak areas, as they are likely to show up on the exam.

6. **Get Plenty of Rest:**
   - The night before the exam, make sure to get a good night's sleep. Your brain needs rest to function at its best, and staying up all night cramming is likely to hurt your performance. On the day of the exam, eat a healthy meal and arrive early to the exam room.

Overcoming Common Challenges in College Algebra

Many students face challenges in college algebra, whether it's difficulty understanding the material, managing time, or dealing with math anxiety. Here are some common challenges and how to overcome them:

1. **Math Anxiety:**
   - Math anxiety is a real issue for many students, and it can affect your performance in algebra. To overcome math anxiety, practice relaxation techniques, stay positive, and focus on understanding the material rather than worrying about grades.

2. **Time Management:**
   - Balancing college algebra with other courses and responsibilities can be tough. To manage your time effectively, create a study schedule, prioritize your tasks, and avoid procrastination.

3. **Difficulty with Word Problems:**
   - Word problems can be especially challenging in algebra because they require you to translate real-world scenarios into mathematical equations. To improve your skills with word problems, practice breaking down the problem into smaller parts, identifying key information, and setting up the correct equation.

4. **Struggling with New Concepts:**
   - If you find yourself struggling with new concepts, don't give up. Review your notes, rewatch lectures, and practice problems until you feel more comfortable. If necessary, seek help from your instructor or a tutor.

OUTRO

Passing college algebra may seem daunting, but with the right approach, anyone can succeed. By building a strong foundation in the key concepts, developing effective study habits, and preparing strategically for exams, you can overcome challenges and achieve your goals. Remember, algebra is not just about numbers, it's about problem-solving, critical thinking, and logical reasoning. With persistence, practice, and a positive mindset, you'll be well on your way to passing college algebra and opening the door to further academic and career opportunities.